STEAM CULTIVATION

IN ENGLAND.

A PAPER READ BY

J. H. VAN ALEN

BEFORE THE

Farmers' Club of the American Institute.

NEW YORK:
EVENING POST STEAM PRESSES, 41 NASSAU STREET, COR. LIBERTY.
1872.

STEAM CULTIVATION IN ENGLAND.

A PAPER READ BY

J. H. VAN ALEN

BEFORE THE

FARMERS' CLUB OF THE AMERICAN INSTITUTE.

I am here to-day to tell you what I saw and heard of steam cultivation during my recent visit to England. It is not my purpose to describe the mechanical construction of steam cultivating machinery. I propose rather to consider the practical results which have been obtained by the use of such machinery, so as to enable you to form an intelligent opinion as to its adaptability to your own farms.

STEAM-POWER COMPARED WITH THAT OF HORSES.

Regarding the application of steam-power as a pulling medium, there can be no discussion. That question is settled by experience. Where steam-power can be conveniently applied, there is no doubt that it can be done at one-tenth the cost of horse-labor. A steam-engine set in position, drawing a wire-rope, will pull at a cost of 12 cents per horse-power a day. If such be the case—and I am told it can be proved by numerous cases in collieries in England, where wire-rope is now being increasingly used as a means of conveying the requisite power—there can be no dispute as to its efficiency. With practical men, it therefore resolves itself into a question of application, and that this application can be satisfactorily effected by agriculturists there can now be no doubt. The draft of a horse can be, and is, carried continuously for ten hours in English farms at a cost of 6d. sterling per acre. Steam plowing is largely a matter of pulling; and, provided the necessary pulling power can be obtained, the next consideration is to arrange the conveyance of this power in such a way as shall be most suitable for the use of the agriculturist.

PRESENT STATUS OF STEAM CULTIVATION IN ENGLAND.

Regarding steam-plowing as it is, its present position is somewhat peculiar. It is an admitted fact that in steam-cultivated land the crops are very materially increased where judicious management has obtained, and those farmers who devote a reasonable amount of thought and energy to the work are succeeding and making plenty of money. One wealthy and intelligent gentleman told me that his farm in the County of Notts had doubled its product and net profit since his introduction of steam-plowing.

One of the great advantages of cultivation by steam is that in all cases the drainage is thereby very greatly improved; and it is a noteworthy fact that, in many instances, the use of pipes for draining purposes has ceased, the steam-cultivator having been found to be all that was needed to put the land in the same condition.

It is, too, another noteworthy fact that in many counties where the steam-plow companies have experienced great difficulty in getting the farmers to use their machines, the next year the applications have been greater than they could supply. Indeed, I was told that it was rarely the case that a farmer, having a field steam-cultivated one season, would not have three or four times as much land done the year after. From this, we may infer that, as a rule, the farmer is quite satisfied with the character of the work.

Until the last two years, steam-plows have been used but little for any except the heavy operations. This has been found to be a mistaken policy. Let it be borne in mind that in horse-plowing each horse-foot mark which is put upon the land has to be taken out again, the land having, first of all, been trodden down in a very uneven way, part of it being soft and part of it hard. It was found to be exceedingly desirable, with a view to economy, that the soil during all cultivating operations should be entirely free from animal foot-prints. As an illustration of this point, I was told that it is very common for gardeners to have their harrows pulled by men rather than have animals treading and hardening the ground. If the animals are kept off the ground and the land plowed in proper season, one-half of the present cultivation has been found quite sufficient. No clod-crushers or mechanical means of breaking the soil will be required.

It is often said that deep cultivation will not do for all crops. Now, I am entirely of opinion that deep cultivation is unnecessary more than once in every rotation; but this cannot be accomplished unless every operation is done by some means which do not involve any trampling or treading. I am confident, from the testimony of experienced men, that the cultivation of the land in England, where steam-plowing operations are thoroughly carried out, does not cost more than eight shillings sterling per acre per annum. The operations can be done at the exact time the farmer requires them, or when his land is in the most suitable state for their being done ; no extra expenses will be incurred when a wet day occurs ; and it will be possible to keep the farming expenses within the estimate which has been made beforehand.

The crops will be increased in quantity, and improved in quality, in consequence of the cultivation of the land being effected at the right time and in the right condition.

ADVANTAGES TO BE GAINED.

The state of the land where steam-power has been applied for a considerable time is worthy of thoughtful consideration. But few people are prepared to credit the change which has taken place in a piece of soil where the spade has been employed as a means of cultivation for some years.

Suppose a house built on the very poorest soil in the country ; surround that house with a garden. Now observe as soon as the garden changes its color it becomes easy to cultivate. This, perhaps, you suggest, is owing to manure. Not entirely so. Of course it is partly owing to the manure, but the greatest advantage the soil derives is from the roots of the plants conveying the atmospheric elements and influences down into the soil; and hence it follows that, though no manure, except the artificial kinds, be applied, yet if vegetables be planted on it, that ground will become quite black and rich-looking. If spade husbandry has the advantages which I claim for it (and which few tillers of the soil will dispute), steam cultivation has similar advantages, with this important addition, viz., that in the well-timed operations of steam cultivation the work is done with much more regularity, with much greater rapidity, at a lower depth, and without any treading whatever. The great point with the soil is to treat it when it is in a proper state for being operated upon, and never, under any consideration, when it is raw or wet. By way of familiarly illustrating this remark, I may venture to observe that most of you have occasionally plowed down a little snow. In so doing, I have no doubt you have noticed, after the plowing, the lines made by the plowed-down snow. If such is the case, then I hold that turning land over when only slightly wet must produce the same result in a proportional degree.

The temperature of steam-plowed or dug land is always materially raised, and it will keep much nearer one point than other land will on which horses have been trampling and treading, the more so if this has been done in wet weather.

It is an important fact that, other things being equal, any land plowed by steam dries much quicker than any other land not so cultivated. This is mainly caused by the rapid action of the implement in passing through, and also in tearing the subsoil below where ordinary plows had previously penetrated, so tbat the water can now descend into the drains.

As already intimated, the depth of cultivation is a very important consideration. By deep cultivation, of course, I don't mean simply turning over. I mean that the land be thoroughly loosened, and that provision be made to retain enough of the moisture and wet at the bottom, and in such a state as to allow any superfluous water which may fall to pass away. The advantages derived from such deep cultivation are that the surface land never retains more moisture than it requires, and, when dry weather comes, the spongy subsoil keeps it until the plant demands it, thus entirely preventing two injurious extremes.

The subject of dealing with the land may be further illustrated by a passing reference to one of the common laws of health. A man cannot continually be in the best hygienic condition unless he pays some attention to the temperature of the atmosphere by which he is surrounded, and in which "he lives and moves and has his being," and unless by lighter and heavier clothing he endeavors to secure something like an equality of temperature in the heat of summer and the cold of winter. This point is of much more importance in its bearing on the growth of agricultural produce than most farmers imagine. Plants require as even a temperature as it is possible to secure, and the only means of effecting this are deep cultivation and thorough drainage. With respect to manuring from the atmosphere, it has been well observed by an eminent agricultural authority, to whom I am deeply indebted for much of what I know on the subject of this paper, and from whom I have largely drawn for my facts and their logical sequences, that soils, if put into proper condition, are able to imbibe a vast amount of manure from the atmosphere—much larger than most persons imagine. They will absorb from the air ammonia, a very large quantity of which substance the atmosphere gives to the land in the form of rain. This at once raises the subject of draining, for if land is clogged with water to the surface, it is unable to benefit by the valuable manures which descend in the rain, which, instead of soaking through the ground, is compelled to run off the surface without giving half its manuring value to the crop.

Again: undrained land is unable to extract the manures from the air, for this process is only carried on by reason of its porosity; and, therefore, if the water cannot pass readily away, the ground remains full, the air cannot pass through it, and thus the ground derives no benefit from it.

This power possessed by the soil of manuring itself from the air constitutes the real value of fallowing, and is greatly assisted by deep cultivation. Another point which the English farmer deems it all-essential to study is never to touch the land except when it is in the best state for the work to be done. If plowed wet, it has been found that the temperature of the land will be lowered for the entire season.

COST OF STEAM CULTIVATING MACHINERY.

The Single Cylinder Traction Engines range from 8-horse to 20-horse power, and without the implements, cost from £657 to £2,495.

The cost of the 12 and 14-horse-power, which will be found to be the best adapted to our purposes, is £1,600 and £1,670, respectively. This includes all the implements—plow, cultivator, harrow, water-cart, &c.

COST OF PLOWING PER ACRE.

This is difficult to get at, because it depends so much on the character of the land, the state it is in at the time, wages, coal, weather, &c. The Northumberland Steam Cultivation Company contract to plow, dig, cultivate, and harrow at the following prices, subject to coal and water being found by the farmer at all times when required, and assistance given in moving to the next place when the work is completed.

Plowing barley and wheat land, not exceeding 7 inches in depth, 10s. to 12s., or $2.50 to $3 per acre.

Stubbles, not less than 8 inches, nor exceeding 12 in depth, $3 to $3.75.

Cultivating.—Heavy cultivator, not exceeding 10 inches in depth, $1.88 to $2.50 per acre; heavy cultivator, exceeding 10 inches in depth, $2.50 to $3.50 per acre; heavy cultivator, second time over, $1 to $1 88 per acre. Light cultivator, $1 to $1.88 per acre; light cultivator, second time over, 38 to 50 cents.

Harrowing.—Single time, 62½ cents per acre; double time, $1 per acre.

The total cost then of plowing, cultivating, and harrowing per acre would be from $5.38 to $6.50.

RATE OF ACRES PER HOUR WHICH CAN BE PLOWED.

This rate depends so much upon the same considerations which affect the cost that I can only approximate it. I should say that, on the average, an acre of very stiff clay soil can be plowed in an hour, though in competing trials, where the land was in the best possible state, and all the other conditions were favorable, the rate per hour made by the largest sized traction engine, viz., 20-horse-power, has been increased to 4½ acres. In one of these trials, the number of revolutions was frequently from 250 to 350 a minute.

COMPETING TRIALS OF STEAM CULTIVATING MACHINERY.

In July of last year a trial was made, lasting several days, of competing steam cultivating machinery, a description of which cannot fail to interest the intelligent farmers of our country. I therefore offer no excuse for drawing largely upon the report made by a clever expert. The trial took place at Wolverhampton, at the Show of the Royal Agricultural Society of England. Six different sets of apparatus were put to work by the judges, viz.: 20-horse-power traction engines, 12-horse-power traction engines, 12-horse-power, with clip drum and traveling anchor, and 8-horse-power, with two winding drums and traveling anchor. All the engines were self-moving along the headland, and all constructed with single cylinders and reversing gear. The implements used were cultivators, which are turned round at the ends of the work, excepting in the case of the clip-drum set of apparatus, which worked a balance-cultivator fitted with compensating gear for holding taut the tail-rope.

Some of the marvels of steam cultivation are exemplified in the following feats: The pair of 12-horse engines, with the implements and all tackle belonging to them, traveled about half a mile, passing through two awkward gateways, and took up position in the field, ready to start work, in the short space of 15 minutes; and after completing the plot of work, the whole apparatus was taken up, and in readiness for the journey out of the field, in only eight minutes. No help was wanted from horses, except the necessary attendance of a water-cart. The 20-horse-power set of machinery, equally manageable, unless it be upon very wet land, is an illustration of the extraordinary advance made in steam tillage since the early trials of twelve or fifteen years ago, when every effort was bestowed upon utilizing, if possible, the threshing machines which were already in the farmers' hands. At 100 pounds pressure, and 130 revolutions per minute, the engines are nominally of 20-horse-power each; but in work that pressure was greatly exceeded; the speed sometimes was fully 360 revolutions per minute, and the actual force exerted may be estimated at 120-horse-power for each engine.

Here, then, is a triumph of agricultural engineering—a motive-power comparable to that of the great engines which drive the heavy machinery in works and mills, transporting itself with the utmost ease over farm roads and unlevel fields, and executing tillage operations with astonishing rapidity and effect. Driving a nine-tined cultivator at about eight inches depth, through a very foul clover lea, and tearing and shattering the sandy loam soil by the pace of the instrument (which was sometimes six or even seven miles per hour), the pair of engines finished a three-acre plot in 43 minutes, being at the rate of fifty acres in an autumn day of twelve hours.

The Fowler pair of 12-horse engines, with a nine-tined cultivator, at 8½ inches depth in a clover lea, performed at the rate of forty acres in a day of twelve hours.

The single 12-horse-power engine, with a clip-drum and anchorage, driving a seven-tined cultivator seven inches deep, in similar land, worked at the rate of twenty-three acres per day of twelve hours, and the single 8-horse-power engine, with two winding drums, anchorage, and a five-tined cultivator, making exceedingly good work eight inches deep in the same description of soil, performed at the rate of twenty acres in a day of twelve hours.

These areas may be compared with the six to ten acres per day which we so commonly hear of as attainable by single-engine apparatus upon farms, and it may also be well to reflect how the existence and complete success of giant engines for field husbandry point to a rapid extension of the contract system, which alone seems likely to carry steam culture within reach of innumerable small farmers. For, if the concentration of such a force as 240-horse-power in the hands of a couple of engine-drivers and a plow-man, can cheaply dig or plow or smash up for the small or medium farmer in one day a field that would have occupied his teams for a week, no amount of inertness among husbandmen, or fondness for antiquated and obstructive modes of tenure among their betters, can prevent the march of this mechanical revolution through the light-land as well as clay-land districts of Great Britain.

As a set of tackle, which may require 20 to 30 acres daily for its profitable employment, and this area of fresh ground, for say 100 days in a year, will demand 2,000 or 3,000 acres of work annually, it appears that either farming on a great scale, or an extraordinary development of the contract system, or both, are necessary results of the introduction of this class of steam-tilling machinery.

To work the apparatus of the 12-horse-power double-cylinder portable engine, stationary winding windlass and anchored pulleys on the roundabout system, requires four men, beside porter boys; but the system involves very little labor or loss of time in removals. In fact, a whole farm may be cultivated from a portable engine, stationed at a central point, where there may be a supply of water without carting, or even from a fixed engine erected inside the farm buildings; or again, from a water-wheel or a wind-mill, the waste of motive-power in transmission for half a mile or more being apparently not a very serious consideration.

Among the prizes was a cup of the value of £100, offered by Lord Vernon "for the best combination of machinery for the cultivation of the soil by steam-power, the cost of which shall not exceed £700, the engine to be locomotive, and adapted for threshing and other farm purposes." The prize was awarded to Messrs. John Fowler & Co., of Leeds. This machinery consisted of a single 8-horse-power movable engine, with two winding drums working a wire rope in connection with a self-moving disc anchor pulley at the other end of the field, which for the rope supplied may be 600 yards long. The price, with combined 4-furrow plow, digger and cultivator, is £698.

A silver medal was also awarded Messrs. John Fowler & Co. for their "ditching plow," an implement which, with surprising power, excavates an open drain of the dimensions of two feet wide at the top, ten inches wide at the bottom, and 24 inches deep, and at the rate of about a mile length per hour. This implement is of special value for ditching in sugar plantations, and is already in use both in Egypt and the West Indies.

I cannot resist the temptation to transcribe here a spirited sketch of a trial of machinery which took place at Wolverhampton. For me it has a charm far greater than any description of a field day at the Derby. There were seven entries in class 1, but all the exhibitors did not put in an appearance. Lots were drawn the preceding evening for order of turns, when Messrs. Barrows & Stewart came first; Messrs. Howard, second; Messrs. Fowler, third; and the Ravensthorpe Company, fourth; but the other competitors not being quite ready, it was agreed that Messrs. Fowler should take the first turn. Every man was at his post, from the proprietors to the stokers, and the discipline observed seemed to have a military precision and strictness about it. The coal and oil having been issued, two 20-horse engines and a 13-tined cultivator started from the depot field exactly at 10 o'clock for the trial field, a distance of about a quarter of a mile. The field was reached in six minutes, but in passing over it, the engine wheels slipping in wet soil, additional grips had to be placed upon them. In half an hour the tackle was fixed, the engines being placed opposite each other on the head-lands. The field looked like clover stubble, but it displayed such a large and miscellaneous crop of twitch, docks, and other vegetable pests, that it was difficult to say what it really was. The cultivator started at a spanking pace, but before it had crossed the field, and returned, it was found that the matted weeds clogged between the wheels and the outermost tines. These latter were accordingly removed, and thenceforward the work went on swimmingly. The pace was tremendous—about seven miles an hour—and the cultivator, as it tore its way through the soil, was tossed about like an open boat in a gale of wind, the resemblance to a boat being heightened by the manner in which the tines scattered the soil, and by the steersman at the stern. The cultivator penetrated to a depth of about 7½ inches, and the plot of three acres was finished in 41 minutes, or at the rate of 54 acres per day of 12 hours. A portion of the soil having been removed, it was found that the bottom had been left level and the weighing of a square yard of the comminuted soil gave the result of 6 cwt. and 5 lbs. The following are particulars of the engine with which this work was accomplished:

Diameter of single cylinder, 13 inches.
Stroke of single cylinder, 14 inches.
Total heating surfaces, 278 square feet.
Average steam pressure, 100 lbs.
Number of revolutions, 130.

Speed of plowing rope, $2\frac{3}{4}$ mile.

Speed of road motion, 2 miles.

Diameter of driving-wheels, 6 feet 6 inches.

Breadth of driving-wheels, 22 inches.

The weight of the engine, complete, and in working order, is 17 tons

After this Messrs. Fowler set to work in another field their 12-horse double set, with a 9-tine cultivator. With this tackle three acres were got over without a hitch, in a minute or two less than an hour, and the soil was broken up to a depth which appeared to vary from $6\frac{1}{2}$ to 9 inches. At the close of this trial, a capital illustration was afforded of the degree of smartness to which Messrs. Fowlers' people have been trained. From the time of completing its task to the re-entry of the whole of the double set into the depot field, a distance of nearly a third of a mile, but eight minutes were consumed.

Messrs. Fowlers' third set consisted of the clip-drum tackle, with a 12-horse-power engine. The weight of the whole of the machinery is under ten tons, this comparative lightness being obtained because the boiler and working parts are made of steel. In this system the engine, instead of having a coiling-drum beneath it, has its drum with a single V groove, composed of a great number of movable clips, which grip the rope so effectually as to prevent it from slipping. On the opposite head-land a self-moving anchor on six-disc cutting wheels, travels along as the work proceeds. The implement, which in this case was a 7-tined balance cultivator, is provided with a quantity of wire-rope wound on two drums, by which the rope in use can be shortened or lengthened according to the necessities of the work as it proceeds. The cultivator or implement is hauled between the engine and anchor and back by an endless wire-rope, passing round the clip-drum of the engine and the sheaf of the anchor. Three men and two boys worked this tackle. The ground was broken up to the depth of about seven inches, and the three acres were completed in about an hour and a half. The tackle was 38 minutes in moving from one field to the other and setting down to work.

The next day some splendid steam-plowing was done by a large variety of implements entered by different manufacturers. The exhibitors were allowed to work as slowly as they pleased, so that the furrows were laid with mathematical precision.

In contrast to this fine work of the ordinary three, four and five-furrow plow was the work of one of Fowlers' two-furrow deep plows, which penetrated to a great depth, and in its course came upon a large number of masses of sandstone, some of which, it is conjectured, weighed more than a hundred weight. These obstacles the plow tossed about and threw aside as though they had been so many pebbles. The scene at Hopton the next day was very animated. The tests were virtually completed toward 1 o'clock, when orders were issued to suspend work pending the arrival of members of the Council and other distinguished visitors from Wolverhampton. There was consequently a lull for about an hour, at the end of which time what might be regarded as a sort of parade view took place. Exhibitors were left at liberty to choose their own implements and class of work, and it was evident that all exerted themselves to the utmost to make a favorable impression on the eminent agriculturists and engineers who had by this time congregated. The high pitch of discipline attained by Messrs. Fowlers' staff was again illustrated in a remarkable manner, one set of implements after another being worked rapidly without the slightest hitch or miscarriage. The novelty of the day was a ditching or trenching machine, which at one stroke cuts a trench two feet deep, and with the soil thrown up, four or five feet wide. The cutting part is a spade or flat share seven inches in width, for cutting the bottom of the ditch. It is carried by two strong coulters, which cut the sides; a third coulter behind splits the cut earth in two, and it is then lifted up an inclined plane and pushed away at some distance on either side of the trench, with a motion resembling that of a snow-plow. In fact, the solid earth was cut and tossed aside as unceremoniously as if it had been a feather snow-drift, to the amazement and admiration of the spectators. The frame, which is of strong angle iron, is carried on two main wheels, five feet six inches high, and on two smaller ones in front. The common axle of the hind wheels is cranked, and provided with a segment similar to the one used in the turning cultivator.

This ditching-machine was intended for the drainage of the cotton and cane low lands of Louisiana, and it will be effective for irrigation where it is employed on a large scale.

It is claimed by those who have thoroughly tested steam cultivation that its advantages are:

Firstly, It deepens the soil. An immense area of England has never been plowed to a greater depth than four or five inches, and the consequence is that the land is impoverished by the trampling of the horses year after year, and rich subsoils, instead of being rendered fertile by exposure to atmospheric influences, are hardened, until neither roots nor water will penetrate. Deep cultivation is, in fact, the exception, but its importance can scarcely be overrated.

Secondly, There is the advantage of increased expedition and autumn cultivation. Time is money in the autumn, and it is directly after harvest—when the sun is hottest, the land is dryest, and the weeds are most easily destroyed—that the land wants breaking up; and this may be done by a steam cultivator, at the rate of from ten to thirty acres per day.

It improves drainage; it does away with the trampling of the horses—and that is not a trifling consideration, when it is remembered that 350,000 hoof-marks (the curious or skeptical may easily verify this statement) are imprinted on every acre of land plowed by horses. Again, fewer horses are required on a farm; and every horse dispensed with means a saving of from £40 to £50 a year.

If steam-plowing be continued for a few years, the cost of preparing the land is brought to a minimum. Mr. William Smith of Woolston, than whom there is no higher practical authority, asserts that the total coast of preparing his seed-bed for wheat, including every expense, is only 7s. 6d. per acre. This, the writer adds, may strike some minds as almost incredible; but it must be remembered that Mr. Smith's farm has been under steam cultivation for many years.

Lastly, Steam cultivation improves the stȧtus of the agricultural laborer by increasing his intelligence and enhancing his value to his employer.

ILLUSTRATION OF THE RESULTS OF STEAM CULTIVATION.

Several objections have been urged against the system, but for answer thereto, as well as for information on a variety of collateral topics, I cannot do better than refer you to an able pamphlet by Mr. Isaac Robinson of Wisbech. From the last page of his pamphlet, I am tempted to give you the following striking illustration of the results of steam cultivation and unrestricted farming: In 1861, Mr. John Prout bought 450 acres of clay land near Sawbridgeworth, Hertfordshire, at £35 per acre. He grubbed six and a half miles of useless hedgerows, threw the land into square plots divided by engine roads, drained it, and gradually deepened the soil by steam cultivation. He has sold all his crops off the land for the last seven years, has used hardly any manure, except artificial, and, so far from the land becoming impoverished, its fertility, and consequently his crops, has increasėd year by year, while his expenses have, year by year, diminished. Four horses will now do the whole of his farm-work. Last year's crops he sold, as they stood, for £5,330, and, after paying all expenses, and deducting £2 per acre for rent, and 10 shillings for rates, &c., his tenant profit for the last three years averaged £1,150 per annum. A degree of success like this could not be achieved by a tenant-farmer unless he and his landlord pulled together—the latter granting a long lease, or an equitable and liberally conceived tenant-right; the former bringing to his business energy, education and capital.

But I must bring to a close a paper for which I claim little more credit than is due to a compilation. I have tried to show you that, through the convenient and untiring agency of steam, the farmer's work can be done at less expense—done, too, so thoroughly that the produce of each acre is nearly double—for the land is not only plowed, but at the same time fairly drained; fewer horses are required, a great saving of time is effected, the condition of the farm-laborer is improved, and agriculture is pursued as an interesting and practical art, as well as a promising and profitable business.

I shall feel that my object has been fully accomplished if the facts I have collected shall awaken such an interest in steam cultivation as shall cause its general adoption in a country where, by reason of the fact that in its great agricultural districts the planting season is short and the laborers are few, the steam-plow ought to find its heartiest welcome. Our mighty West is already the granary of our country; it is a leading resource for breadstuffs of the British isles. Its vast, level, fertile, stoneless, treeless prairies are especially adapted to steam cultivation. Its high prices of labor demand the largest possible economy of human muscle and human supervision. Its immense grain-fields can be adequately and reasonably tilled only by steam. Harness steam to her plows, and the West may rapidly double and quadruple her harvests, feeding the hungry of this country and Europe while she enriches her farmers by the munificence and certainty of their annual harvests.

www.ingramcontent.com/pod-product-compliance
Lightning Source LLC
LaVergne TN
LVHW020644110826
845149LV00004B/1346

* 9 7 8 1 4 1 8 1 9 0 5 7 6 *